# Space Rescue Systems

## In the Context of International Laws

## Other Works by the Authors

**By Austin Mardon**

*A Conspectus of the Contribution of Herodotus to the Development of Geographical Thought* 1990;
*International Law and Space Rescue Systems* 1991;
*Kensington Stone and Other Essays* 1991;
*A Transient in Whirl* 1991;
*Alone Against the Revolution* 1996;
*Political Networks in Alberta 1905-1992* 2002;
*7 days in Moscow* 2005;
*The Contribution of Geography to the Recovery of Antarctic Meteorites* 2005

**By Ernest Mardon**

*Narrative Unity of the Cursor Mundi* 1967;
*The Founding Faculty of the University of Lethbridge* 1968;
*Place Names of Southern Alberta* 1970;
*The Conflict Between the Individual & Society in the Plays of James Bridie* 1971;
*Who's Who in Federal Politics from Alberta Ridings* 1972;
*English Studies at Canadian Universities* 1972;
*Community Names of Alberta* 1975

**By Austin Mardon & Ernest Mardon**

*Alberta Judicial Biographical Dictionary* 1990;
*Alberta Ethnic Mormon Politicians* 1991;
*Alberta Ethnic German Politicians* 1991;
*When Kitty Met the Ghost* 1991;
*Down & Out & On the Run in Moscow* 1991;
*The Girl Who Could Walk Through Walls* 1991;
*Alberta Mormon Politicians* 1992;
*Alberta General Election Returns & Subsequent Byelections* 1882-1992 1993;
*Edmonton Political Biography* 1994;
*Alberta Political Biographical Dictionary* 1994;
*Alberta Executive Council* 1905-1990 1994;
*Early Christian Saints* 1997;
*Later Christian Saints for Children* 1997;
*Many Christian Saints for Children* 1997;
*Childhood Memories & Legends of Christmas Past* 1998;
*Community Names of Alberta* 1999;
*Men of Dawn* 1999;
*United Farmers of Alberta* 1999;
*The Genealogy of the Mardon Family* 2000;
*Alberta Catholic Politicians 2000; Alberta Anglican Politicians* 2001;
*Liberal Politicians in Alberta* 1905-1992 2002;
*What's in a Name?* 2002;
*Edmonton Members of the Legislature* 2004;
*Senators and Members of the House of Commons from Edmonton* 2004;
*Edmonton Municipal Politicians* 2005;
*Alberta Francophone Politicians* 2007;
*English Comets over the Medieval Period* 2008

# Space Rescue Systems

## In The Context Of International Laws

Austin A. Mardon, Catherine Mardon, Brenda Ball,

Jocelyn Best, & Zack Best

Edited by Lawrence Dommer

Golden Meteorite Press

Edmonton

A Golden Meteorite Press Book



Supported by a grant from the Antarctic Institute of Canada.

Cover design Lawrence Dommer, 2011
Published by Golden Meteorite Press
126 Kingsway Garden
Post Office Box 34181,
Edmonton, Alberta, CANADA.
T5G 3G4Telephone: 780-378-0063

**ISBN** 978-1-897472-21-7

**Library and Archives Canada Cataloguing in Publication**

**Library and Archives Canada Cataloguing in Publication**

**Space rescue systems in the context of international laws / Austin A. Mardon ... [et al.] ; edited by Lawrence Dommer.**

**Includes bibliographical references.**
**ISBN 978-1-897472-21-7**

**1. Space rescue operations. 2. Space law. I. Mardon, Austin A. (Austin Albert)**
**II. Dommer, Lawrence**

**KZD4326.S62 2011 341.4'7 C2011-900785-1**

**Dedicated**
**To**
**Sumi**

# Space Rescue Systems in the Context of International Laws

Abstract:

The Agreement on the Rescue of Astronauts, the Return of Astronauts, and Objects Launched into Outer Space of 1968 (Assistance Agreement) is an international treaty that could become a legal test bed for the elaboration, strengthening, and expansion of the space law regime. A revised and expanded Assistance Agreement could be the result of a proactive stance to introduce standardization of human and machine rescue systems. It could also strengthen general attempts to reduce man-made environment environments. The support of rescue capabilities and rescue services for all space-faring peoples strengthens the central aspect of the outer space legal regime. That central aspect is the provision that space is the "province of all mankind." The "province of all mankind" general principle shows that humanitarian interests play an important part in our legal perspectives off of Earth.

## 1. Introduction.

The Agreement on the Rescue of Astronauts, the Return of Astronauts, and Objects Launched into Outer Space of 1968,[1] hereafter referred to as the Assistance Agreement,[2] encompasses the full breadth of concepts that were formally the de jure purpose of international space law. The Multilateral Treaty on Principles Governing the Activities of States in the Exploration and Use of Outer Space, including the Moon and other Celestial Bodies, hereafter referred to as the Outer Space Treaty,[3] codified the first international treaty reference to the rescue and aid of astronauts/cosmonauts and, as such, forms the basis for the later development of the Assistance Agreement.[4] This same Assistance Agreement[5] supports and strengthens the international legal framework of law in space. There was a distinct evolution from Article V in the Outer Space Treaty[6] to the Assistance Agreement.[7,8] The same fact, while applying to all of the treaties made in relation to humanity's activities in outer space, is especially true in the case of the Assistance Agreement.[9,10]

From the dawn of our existence in the misty reaches of the Pleistocene epoch, we have recognized the waylaid traveler in need of help. Many early societies have ritualized aspects of their culture directly relating to the assistance of strangers that were shipwrecked or somehow put by fate in distress. Maritime history is rife with examples of shipwrecked voyagers that have been assisted under established maritime international law, which provides for the rescue of sailors and passengers.[11] The Assistance Agreement[12] is analogous to the earlier treatments of distressed seamen.[13] As such, it shows the continuity of thought from the seafaring times of human travel to the space-faring days of human travel.[14] This same universal custom of assistance to the waylaid stranger was encompassed in the conceptual framework of the Assistance Agreement.[15] The space law merely formalizes what is self-enlightened interest. At very early points in the space programs, the American government was pushing technical efforts to develop rescue capabilities,[16] but due to the Cold War,[17] it was not possible to expand on the good intentions of integrating international space rescue capabilities.[18,19] On Earth, international space-based disaster relief technologies have been standardized and utilized on a global scale.[20] These same expressions of the humanitarian wishes for distressed "envoys of mankind"[21] should exist today and could be achieved by major revisions in the text of the Assistance Agreement[22]
.

The engineering elements of the rescue of distressed astronauts are the mechanism by which any obligation would be implemented by states that are signatories of the Assistance Agreement.[23] A precursor statement by the legal authority Professor E. Pepin from the early 1960s states "preparation of a convention…

would benefit the safety of the future of the circulation of man in space." [24] This shows the underlying intent that predates any of the space treaties. The technical and fiscal factors of rescue of distressed astronauts in space have prevented a full rescue capability from developing until the 1990s.[25] An American rescue capability was described in 1963 but was not pursued:

> Development was not pursued, however, because planners realized that successful space rescue would require an ambulance spacecraft to be available for immediate launching, and maintaining such a capability was deemed impracticable at the time.[26]

The lack of any national or international rescue capability persists into the 1990s. The international legal regime, rather than inhibiting activities in space, is and can be a proactive technique that can push technology rather than merely react to it.[27] This would mean that the development of technology related to human factors would have an internationalist safety focus[28,29] along with certain reactive elements. A simple, international standardization of spacecraft components that directly relate to rescue capability would suffice in the interim until dedicated rescue craft are developed. The standardization of spacecraft components directly related to rescue capability would result in vehicles having the potential for double-tasking as rescue vehicles.[30]

## 2. Article V of the Outer Space Treaty[31] and its Bearing on the Aid of Distressed Spacefarers.

There is a historical and contemporary linkage between the Outer Space Treaty[32] and the Assistance Agreement.[33] Debate as to the mechanisms of how the treaty fits into the larger scheme of the international legal regime is ongoing.[34] Due to a number of factors, the legal regime has been in a state of flux as far as mechanisms of implementation and interpretation of these legalities.[35,36] The treaty scheme that predates the Assistance Agreement[37] should always be referenced as this strengthens the international legal regimes that all space treaties will operate under when they are in force. It becomes self-evident upon reading the first sentence of Article V of the Outer Space Treaty[38] that the references to uncontrolled landing and a sea ditching of the incoming spacecraft are a direct result of the re-entry technology of the late 1960s[39] as indicated in the following section:

> State Parties to the treaty shall regard astronauts as envoys of Mankind in Outer Space and shall render to them all possible assistance in the event of accident, distress, or emergency landing on the territory of another State Party or on the high seas.[40]

The general principle of astronauts as the "envoys of mankind"[41] does not vary with a change in technological context of the space environment. Under any new regime in space, a revision or the addition of amendments to the Assistance Agreement,[42] the Outer Space Treaty[43] will have to be implemented.

The Article V of the Outer Space Treaty[44] set the stage for the further expansion of it, which comprised the intent of the text of the Assistance Agreement.[45] The negotiations for the Outer Space Treaty[46] encouraged this by laying the broader framework of the international law outside of the Earth planetary environment. The brevity of Article V,[47] which covers the disposal of the distressed astronauts and other craft, indicates the need for clarification and a separate legal treaty to cover all of the pertinent aspects of this issue. This set the stage for the Assistance Agreement[48] to deal exclusively with the issue that was briefly touched upon by Article V of the Outer Space Treaty.[49]

## 3. Analysis of the Text of the Assistance Agreement.[50]

The Assistance Agreement[51] contains ten articles, six of which deal exclusively with unique outer space concepts. These articles will be examined and critiqued. The first part of the Assistance Agreement[52] sets out the framework for the later articles of the Assistance Agreement.[53] Specific sections of the text are in need of revision and are based on the specific technological context of global rescue capabilities in the late 1960s when the Assistance Agreement[54] was composed. The first section of the Assistance Agreement[55] refers to the Outer Space Treaty.[56] The intent of this part of the treaty sets the assumptions and the conceptual framework in which the treaty will be enacted. The preamble of the Assistance Agreement[57] sets the legal precedent.

The text of Article 1, the preamble of the Assistance Agreement,[58] states:

> Each contracting party which receives information or discovers that the personnel of a spacecraft have suffered accident or are experiencing conditions of distress or have made an emergency or unintended landing in territory under its jurisdiction or on the high seas or in any other place not under the jurisdiction of any State shall be immediately returned.[59]

The intent of Article 1 of the Assistance Agreement[60] is to define the basic situations where the treaty shall become activated. This section of the Assistance Agreement[61] leaves the impression that landing difficulties are perceived to be a situation that should be reacted to rather than a space accident that should and can be prevented. In any activity that involves machines with so many parts, there is no question that an accident will eventually result. The most commonly known accidents in space activities are following four accidents: the Apollo 13 accident, the 1985 Challenger accident, the Gemini 8 accident,[62] and the 1988 Soyuz accident.[63]

Article 5 also ties in directly to the Liability convention[64,65] that is part of a different treaty in international space law, while Article 6 deals with organizational matters rather than the technical rescue capabilities of the late 1960s. The remaining sections, from Article 7 through to Article 10 of the Assistance Agreement,[66] discuss the mechanisms by which the treaty will be enacted as well as various diplomatic niceties that do not directly relate to the topics of space rescue and might appear in any number of treaties that are currently on the books. This portion of the Assistance Agreement[67] need not be amended based on the technological developments that have compounded rescue operation scenarios.

## 4. Meaning of the Text of the Assistance Agreement[68] When it Was Written.

The treaty was written in a time when the launching and space-borne technologies were distinctly different than they are today. To exemplify this distinction based on the landing technology affecting the terminology and wording of the treaty is the reference in Article 3 of the Assistance Agreement[69] where it states, "… the personnel of a spacecraft have alighted on the high seas."[70] This is a clear indication of the type of re-entry system that was being used and that extensive communications with the re-entering spacecraft were not the norm when this section was composed. Today, this section is not a concern. What might occur instead is allowing landing privileges on paved airstrips with prior consent.[71] This acknowledges the development of the United States (U.S.) Space Shuttle and the Union of Soviet Socialist Republics (USSR) Buran shuttle system that can land on paved landing strips.

## 4a. Technological Background of Space Systems and Historical Effect on the Assistance Agreement.[72]

The Assistance Agreement[73] is a major act in international law that covers the potential accidental landing of manned spacecraft on foreign soil. This treaty covers both the disposition of the vessel and the astronauts/cosmonauts who are pilots and passengers inside the craft. At least indirectly from the text of the agreement, there seems to be some consideration of the legal status of the astronauts and whether the treaty would allow the astronauts to defect or in some way not return to their country of origin/launch. This matter was settled in the 1960s by supporting the position that the astronauts must be returned to the country of launch.

## 5. Context of the Assistance Agreement[74] Within the Outer Space Treaty[75] Legal Regime.

Humanity's early activity in space led to grandiose, sweeping gestures made by the international community that would not occur in the world's present geopolitical climate. Simply by reinforcing the legal matrix of precedent in international law, any revisions to the Assistance Agreement[76] support the globalization of space. Within the Assistance Agreement,[77] this concept involves standardization of manned and unmanned rescues. The general principle of the "province of all mankind"[78] is strengthened if all countries have equal access to humanitarian rescue under situations of distress.

## 6. Standardization of Manned Vehicles and Stations in Space.

The process of industrialization has been a process of greater uniformity in production techniques and systems in all facets of life. In the Space Station Agreement,[79] the precedent exists for multinational implementation of standards.[80] According to the March 28th, 1990, Tokyo Accord a standard was developed for the Space station:

> The equipment will be standardized so that it can fit interchangeably in any part of the station; despite the fact that the modules are being designed, and built on three different continents... all of the cabinets/racks for the station will house standardized systems... will have minimal cost impact.[81]

The Tokyo Accord, within the limited aspects of the American space station, is an example that could be duplicated at a broader international level with total standardization, especially of rescue supporting technologies.

Law at the national and international level is a distinct factor for development of certain types of technology. That certain technologies are affected by international law is of no question in the modern world. The most distinctively clear example of law affecting the way in which technology is developing is the Nuclear Non-proliferation treaty, which attempts to limit the production of nuclear weapons within nuclear weapon-producing nations and the dispersal of technology to nuclear weapon-capable nations. This technology is limited in its dispersion and spread among the non-nuclear capable nations of the world. Similarly, legal mechanisms could be enacted in the expansion and strengthening of the space rescue concept that underlies the humanitarian motives of both the Outer Space Treaty[82] and the Assistance Agreement.[83] Rather than creating a reactive treaty as the Nuclear Non-proliferation treaty is in the prevention of the dispersal of certain technologies, the legal process could be a proactive mechanism that encourages the spread of technologies that will allow the easy integration of different nations' space-faring systems. With the consultation of all existing space-faring nations in both launch capability and usage of space-based technologies, a potential expansion of the Assistance Agreement[84] could allow a slow, orderly standardization of systems that have a direct bearing on perceived future rescues that would occur in outer space.

## 7. Future Possible Technological and Political Developments and Their Possible Effects on the Evolution of the Text and Meaning of the Assistance Agreement.[85]

The law can affect and change the manner in which societies pursue their impulses to explore and colonize new frontiers. The mechanisms for the management of expanded space transit and the necessity to protect the life and limb of those in transit can either be through existing government agencies or through new supra-national government agencies.[86] During the November, 1989, Annual Congress of the Association for Space Explorers (ASE) held in Riyadh, Saudi Arabia,[87] a resolution was drawn up that specifically dealt with the topic of space rescue:[88]

> The Association of Space Explorers urges each space faring nation to commit to a principle of procuring a universal space rescue capability for all people travelling and residing in space and, furthermore, to implement this principle to the maximum extend technically, operationally, and financially practicable. The association appeals to all space agencies to initiate contacts that would lead to specific solutions to problems associated with universal space rescue. The following topics should be included: 1. The development of international standards to achieve compatibility among orbital systems. 2. Consideration of a quick reaction space rescue capability. 3. The establishment of a border international search and rescue infrastructure to incorporate those requirements unique to manned spacecraft reentry and landing.[89]

The Association of Space Explorers (ASE) has dealt with some of the major concerns that should be faced under a new or expanded legal regime that could deal with the space rescue question.[90] With an environment that is so hostile to our species at any level of interaction with it, we must be cognizant of any social mechanisms that will allow our easy transition to survivable interaction modes with the newly transformed ecumene.

With the current trends towards geopolitical change in Eastern Europe, the potential amount of cooperation is increasing exponentially every day between the West and the Soviet Union.[91]

The increased level of interaction between the superpower blocks will necessitate further development of the text of the Assistance Agreement.[92] This will facilitate the integration of the manned and unmanned systems when they are utilized in space. The implementation of one language usage in orbit and in the direction and control of launch as is the case within the civilian air sector's use of English will expedite rescue missions. If all astronauts/cosmonauts had to learn 2000 words of the other space-faring power's language, this would facilitate the rescue of manned vehicles. During a rescue operation, a lack of understanding of the other crew's language might jeopardize the lives of all parties involved. The refueling of satellites' alignment control jets when this becomes technically feasible might accelerate the standardization of satellite mounts and farings on all unmanned spacecraft. Without invoking the nemesis of technology transfer, it is possible to standardize certain non-critical, safety-applicable areas of spacecraft usage. Power and orbital fuels portals for resupply and refurbishment of on-board inclination fuel supplies could be achieved today if all satellites had standard fuel injection portals which could be refueled by a robotic supply vehicle. This would increase the mean lifetime of satellite operations.[93] In a distress situation, stranded human astronauts could also siphon off this fuel if needed in a rescue scenario.

Existing technical plans for the revision of the space station are varied but would involve some form of a life boat.[94,95] The technical and engineering questions that would be faced by a wide variety of accident and response scenarios in space are being studied currently.[96] No reference has been made as yet to the legal and treaty arrangements that would have to be made to assist the development of internationally-integrated space rescue and assistance capabilities. The U.S. is creating legal arrangements on an ad hoc basis to provide logistical support for potential rescue operations. The prime example of this is the current arrangement with Chile to land the Space Shuttle on Easter Island in case of an emergency.[97] This is an example of the U.S. not developing multilateral treaty systems but developing bilateral systems. As is usual within the U.S. space-faring community, the focus is on the technical solutions, ignoring the fact that a legal regime that created greater integration of rescue capabilities might save on the engineering expenses that will have to be invested. The USSR might already have existing proven rescue and assistance capabilities[98] that, if copied, might save all space-faring nations substantial monetary expense.

## 8. Changes to the Assistance Agreement[99] that Should Occur from 1990 through 2000.

A supplemental agreement to the Assistance Agreement[100] must be updated to cope with the technological, exploration, and geopolitical changes that are being projected from 1990 to 2000. In just one decade from today, three different power blocks will have attained the capability to support ongoing, independent, Low Earth Orbit (LEO), human-manned capability. This could be seen as permanent space station complexes or rotating temporary Earth launched vehicles. The Soviet Union has had, and will continue to have, a lead in permanent human presence in space[101] within the space station context and in the application of modular structures.[102] This could be due to the Soviets having had eight space stations[103] as compared to one American space station. The Shuttle will be the workhorse for the development of a Low Earth Orbit (LEO) infrastructure during this period. Due to the extended period of time it would take to implement,[104] it would not be feasible, under normal conditions, for the United States to envisage rescuing their own astronauts from a ground-launched vehicle in the coming decade.[105] This would mean that the Soviets would have to be asked to rescue any astronauts with their existing rescue capability.[106] As Tass stated, "it [Soviet Manned Maneuvering Unit] was created also to provide cosmonauts with a rescue vehicle in case of a mishap."[107] A replacement U.S. vehicle and any expansion of its shuttle fleet will not happen until at least the end of the 1990's.[108] The potential launch time for a small Soviet manned launch is at the maximum twelve days[109] from a cold start. However, if current speculation is correct, the U.S. will be leading a consortium of countries in the support of a permanently manned Low Earth Orbit (LEO) space station.

The U.S. previously had a manned, permanent space station, Skylab, during the 1970s but decided against a continued support of this facility. The manned station is an option for the current People's Republic of China[110,111] launch program; though as they currently discontinued their manned program, they are unlikely to exploit their capability in this arena. However, during the later periods of human expansion in space, orbital stations will become permanent features in the cosmos and will grow in size and complexity. Since the space station is the major focus of manned activity in space, the questions that are directly related to the preservation of life of those personnel that directly serve on these facilities must be addressed. Rather than being a straight jacket[112] of an unwieldy legal regime, the Assistance Agreement[113] can act as a potent force to save human life in a potential disaster in a Low Earth Orbit (LEO) space station by anticipating rescue scenarios and allowing greater international integration of all station systems that are directly relating to the rescue of distressed astronauts and cosmonauts. New environments and the technologies that are created to survive and function within those environments rear the necessity to create new laws for humans to live by within those societies.

As Hammurabi created new laws for his people to live by when the city emerged as a novel social organization unit, we will need new legal regimes and frameworks to function in space. Legal regimes, especially those covering assistance in times of distress, will be primarily centred around the logistics capabilities and delta-V of rescuing craft. The Assistance Agreement[114] will have to come to grips with the rescue requirements of operating orbiting stations and outposts on the moon and other locations. Caching and prepositioning of emergency survival supplies on various record locations on the moon and other celestial bodies could be construed as proactive measures to aid distressed astronauts/cosmonauts.

## 9. Changes to the Assistance Agreement within the Period of 2001

**through 2025.**

After the turn of the century, the tasks faced by space-faring nations will pull them beyond small habitations in Low Earth Orbit (LEO). The National Commission of Space[115] and the Ride Report[116] both discuss the next step, which is the implementation of manned lunar stations. Countries such as the United Kingdom are considering rescue vehicles and small, man-capable capsules.[117] The British manned space vehicle will be incompatible with the Soviet Station MIR.[118] This habitational situational advance will, in and of itself, create an additional necessity for revision in the Assistance Agreement.[119]

## 10. Changes to the Assistance Agreement[120] within the Period of

**2026 through 2050 and Beyond.**

If our outward movement into space is permanent, then the laws that we formulate, and put on the floor of the legal structures yet to be created, will be with us for a long time. According to Hempsell,[121] who integrated the various U.S. plans relating to the manned development of space, sparse but existing development will occur out to Callisto by 2040.[122] A space rescue capability in this period will be dependent on the state of the art spacecraft that can be developed.[123] The rescue capabilities and costs will involve more than merely going up to Low Earth Orbit (LEO) and carrying stranded astronauts back down. Potential deep spacecraft functioning within the Inner Solar System will be involved. A reasonable timeline would have a small outpost at Callisto by 2050.[124] Callisto could be said to be the beginning of the Outer Solar System.[125] A Martian base will have been set up and be self-sufficient in industrial and living products.[126] This might also apply to the value that might be put on life and the development of a new legal concept of what rescue means. According to the existing Assistance Agreement,[127] all of the inhabitants would be "envoys of mankind"[128] with the status that we assign to that today. The rules that cover individual acts and collective acts on Mars will be little affected by inappropriate laws made on Earth unless consideration for the celestial environment is taken into account.

A larger number of space-capable countries and supra-national organizations, such as the European Space Agency (ESA), will conduct more projects independently of the existing geopolitical power blocks. The ESA is even making plans for a manned, independent space station[129] and independent space rescue systems.[130] A tentative launch date of 2030 is proposed for a European Space Station. That space station would again be completely dependent on an independent design and infrastructure.[131] The ESA would necessitate an independently developed rescue system.[132] Having a legal regime in place that would create the push factor for an extensive use of modular structures that allow complete integration of different source-countries' construction techniques would reduce an ESA space station by several orders of magnitude if it were to be launched by 2030. The U.S. Freedom Space Station agreement is playing a major role in focusing modular construction,[133] safety, and rescue techniques into the arena of awareness of space law. If, at this time, the world has not seriously attempted to reduce technical barriers between rescue and human support systems, this failure will become an extreme detriment to global rescue capabilities. By this time, space will be the "province of all mankind"[134] to a greater extent than it is currently.

Countries who, in 1990, are perceived to have independent orbital launch and support capability will begin to provide these services.[135] One prime example is the Norwegian government whose Andoya Rocket Range[136] is the only European continent launch range.[137] The space community today does not perceive Norway as a major space power, even though it does have a robust program.[138,139] Another potential space-faring block is the Arab Middle Eastern block, as currently represented by the recently announced space capability of Iraq[140] and Israel.[141,142] These tertiary space countries' programs will come on stream with active launch programs and will forestall a plethora of unique rescue systems. It is wise to create legal mechanisms that bring engineers together to solve this problem of the 21st century.

## 11. Proposed Amendments to the Assistance Agreement.[143]

If the Assistance Agreement[144] is not entirely scrapped, then another option might bridge the gap until it becomes politically possible to rewrite the entire agreement. Several timely amendments could be proposed for improvement to the current treaty. They would cover a variety of proactive versus reactive rescue arrangements. Rather than assuming accidents will never happen, one must assume that they will happen and provisions must be made in their eventuality. Several amendments should be added that allow the space-faring nations to increase their interlocking space rescue capability rather than reacting to the symptoms of a problem. While current spacecraft technologies do exist, the use of space will involve certain very specific concerns. In every environment the human race has occupied, we have had to adapt technology for utilization of that environment.

The first proposed amendment to the Assistance Agreement[145] is what might be perceived as a minor matter of flight control systems that would assist directly in an emergency situation. That revision would be the ability of those crewed flights to be able to override the onboard and ground station computer systems and have entirely human control for functions that are directly tied to the rescue of astronauts/cosmonauts in outer space. This would provide for a wide variety of unforeseen emergencies that could be coped with, even if they had to be done entirely by hand. By not stating specific systems, the principles of international space law can be referred to support this amendment, as follows:

> All humanly manned spacecraft should have human override of the control systems of the spacecraft to enable entirely human controlled operations distinctly related to space rescue, assistance, and aid of distressed astronauts and cosmonauts in orbit or outer space.

Since this amendment will only be intended for manned missions and manned craft, it will be comforting to note that a berserk computer accident in space, anticipated in science fiction literature,[146],[147] can be dealt with along with giving the human operators a certain security knowing that if everything goes wrong, they can at least operate the life-preserving systems by hand without any cybernetic assistance. A Soviet incident in 1989 involved the manual override of the computer docking mechanisms,[148] which shows one example of this type of incident. The American Personal Rescue Enclosure (PRE),[149] for example, would be a boon and exhibit the humanitarian nature of this proposal if the designs and technology were released.

The second proposed amendment of the Assistance Agreement states:[150]

> The treaty signatories, realizing the extreme severity of the Outer Space environment and the astronauts are the envoys of all mankind, commit their governments and government sponsored space agencies to support the exchange and development of reliable human support technologies. The geographic area that this will cover is in outer space, with or without gravity or microgravity and any of the

atmospheric and surface environments encountered as of this date and to be encountered within the confines of the solar system on any of the celestial bodies.

This amendment is the strongest proposed amendment to date because it encourages proactive action in the support of the "heritage of all mankind"[151] concept in outer space international law. Under the wider provisions of the Outer Space Treaty,[152] which set down the bedrock principle of space law that many other principles rest on and radiate out from in lattices of mutually support legal precedence, is the concept of space being "the common heritage of mankind."[153] Under this amendment, the newly developing countries will have direct unlimited access to space rescue technology. This will, even in a token manner, support the concept of space being "the province of all mankind"[154] by spreading humanitarian technologies to those countries that might be unable to afford them but would become direct space participants in the future. If universally applicable humanitarian technology is spread, then we might have, ".... The political will to compromise, [that] presents an unparallel opportunity for positive advancement in world affairs."[155] This will, hopefully, immediately support the globalization of space activities.

The third proposed amendment to the Assistance Agreement[156] is simple:

> There shall be standardization of major units of power within the spacecraft system, both manned and unmanned.

If standardization of rescue systems was in place as an Assistance Agreement[157] mechanism, the concept of standardization might have trickle down effect on other areas of multinational space cooperation. The evaluation of the Extra-Vehicular Activity (EVA) technology[158] utilization and the standardization of these technologies has occurred entirely within each space-faring country[159] in complete isolation from all other countries. No attempt has been made to standardize the Extra-Vehicular Activity (EVA) equipment from the time of the United States Mercury program through the Shuttle program and onward to today.[160] The space station has had several embarrassing moments when standardization of some space equipment might have prevented a near diplomatic incident.[161] Since the advent of the American space station, Freedom, the standardization of systems has become the norm in the space station complex.[162] This process is still continuing. If it were decided at the treaty table with all potential space-capable countries, situations such as was faced with the station Freedom could be avoided. If, at some future point, another orbiting mission with the Soviets is contemplated, the technological hurdles will increase with technological complexity. The Soyuz and Apollo docking created the only situation of standard spacecraft docking mechanisms.[163] Even today, a rescue would have to be conducted with Extra Vehicular Activity (EVA) in space suits or some other small personal pressurized device if the Soviets were to rescue Americans or vice versa. If standardization were to have the force of law, as a treaty amendment can give in both countries, it would force the participants to mate and interlock their systems.

The fourth proposed amendment to the Assistance Agreement[164] states:

> A review shall be mandated of technical requirements of international standards of equipment for human support in space every five years or upon the official written request of one of the signatory countries.

This review would necessitate a permanent standing committee under the auspices of the United Nations. Its purpose would be to take the concerns of what should be merely engineering design questions out of the limelight of a continual diplomatic wrangling. The head of the Arianespace, Mous. D'Allest, said during comments in 1988 about the lack of international docking portal standardization that he "found in troubling. [that there were]... no meetings between U.S., Soviet, and European space officials on standardized docking systems."[165] The standardization of docking portals is a prerequisite to the efficient rescue of astronauts stranded in space with egress outside of the confines of the astronauts' pressurized spacecraft.

The fifth proposed amendment to the Assistance Agreement[166] recommends:

> An ongoing committee under the auspices of the Secretary of the United Nations for the supervision and coordination of global standardization of support systems that directly relate to the support and rescue of human life in space.

Under the direction of the United States of America, the space station Freedom is being standardized in the general area of station logistical support.[167] An ongoing committee that deals with this matter would be the more cost effective than dealing with it on an ad hoc basis. A legally-based protocol of agreement might suffice as a beginning point to implement the concept behind this proposed amendment. A topic that is not directly related to rescue but is related to safety is climate and the meteorological effects on the launching[168] and landing of spacecraft. The Shuttle Challenger accident is directly tied-in with weather conditions.

The sixth and last proposed amendment to the Assistance Agreement[169] is:

> A universal verbal and universal non verbal vocabulary for astronaut utilization during a rescue operation of distressed astronauts/cosmonauts will be adopted by all signatory parties. The specific language that the words are to be derived from is the Esperanto language.

All astronauts will then be using a non-native language and will be on an equal footing. The use of this language is for strictly emergency context, not for the purpose of conducting any meaningful conversation. This will mean that all the space faring countries of the world will have to learn the language. This amendment to the Assistance Agreement[170] would relate to the learning of different languages by astronauts and cosmonauts or an independent language such as Esperanto for words that would be used during a rescue operation conducted in space.

Arthur C. Clarke has postulated that the English Language should and will be used in space as the universal tongue.[171] During the Apollo-Soyuz docking operation in the mid-1970s, each of the astronauts learned Russian for 1000 hours and the cosmonauts learned English for 1000 hours. The astronauts and cosmonauts did complain that, at the time, meaningful communication did not happen.

However, learning selected words for emergency situations would be pivotal for clear intent among those performing rescue missions. The use of language is one of the most volatile items of cultural behavior. Esperanto carries no potential accusations that this is a ploy of neo-cultural imperialists and is maintained to be one of the easiest languages to learn. Those who wish to carry their own native tongue into space will be able to, while they will have the assurance that if an accident occurs, any space-faring country will be able to functionally communicate with them in relation to the rescue operation. It could be calamitous if the rescuing party decompressed an airlock without the astronauts on the interior realizing this was about to occur. The required vocabulary used would be distinctly limited in the number of words. Merely 500 words would suffice and would only be intended for use in dire emergencies.

The international standardization of hand signals for use during Extra-Vehicular Activity (EVA) operations has ancillaries on Earth in the standard international signals of combat aircraft[172] and other civilian aircraft that allow communication between pilots without the use of any radio frequency. A set of twenty hand signals would be all that would be required for all astronauts/cosmonauts to learn. They would have a set of universally-accepted hand signals,[173] just as aircraft pilots do.

The various committees could easily function under the United Nations Committee on the Peaceful Uses of Outer Space (COPUOS) or directly under the Secretary-General.[174] Teaching and use could operate under the auspices of the Secretary-General specifically. It is very important to realize that all parties will directly benefit instantaneously by increasing their rescue capabilities, and eventually by saving money through standardization in other areas of space equipment and operations. The standardization that was necessary for the space station Freedom involved extensive and unnecessary international political discussion.[175] After it was tested in the narrow confines of rescue technologies, the concept of standardization could cascade throughout the entire global context of space equipment and activities.

## 12. Structure, Topics, and Proposed Textual Sections to be Included in a New Assistance Agreement[176] that Would Replace the Existing Treaty.

The implementation of a new legal regime must be based on enlightened self-interest of all parties involved with the treaty and its implementation. While compliance to the existing Assistance Agreement[177] is still only voluntary, as the Soviet Union maintained with its reporting of activities in outer space,[178] the current Assistance Agreement[179] is underutilized as a justification for integration of manned and robotic rescue[180,181] systems between the two superpowers.

## 13. Conclusion:

The implementation of a forward-looking Assistance Agreement[182] or amendments to the Assistance Agreement[183] would attempt to use the legal regime to prevent or ameliorate a potential future accident before it happens. It is more cost effective to prevent an accident by reducing procurement costs[184] of up to $1 billion[185] by standardization. The human fatalities or injuries alone that could be potentially avoided are worth the trouble to rewrite or amend the Assistance Agreement.[186,187] Rather than another regime that will spend more money, this treaty could be utilized in such a way to save future monies because of accidents just waiting to happen. International law, specifically the Assistance Agreement,[188] can be a proactive force for the specific improvement of the human condition. Since one of the foundations of international space law is that the astronaut as an individual is an "enjoy of mankind,"[189,190] it is a legal requirement that the launching authorities engage in programs that support the ability to rescue distressed astronauts. This can be specifically interpreted as proactive measures that involve the standardization of farings,[191] grappling devices,[192] respiration gas supply intake valves on space suits,[193] propellant and power systems[194] on Soviet and American Manned Maneuvering Units (MMU),[195,196] and other related devices that would directly support space rescue operations.[197] None of these technologies would compromise national security[198] of any space-faring nation, since they exist in many analog terrestrial environments such as deep sea diving operations and high altitude parachuting. They will only benefit those hapless, stranded astronauts that will eventually have the need to be rescued after an accident in space. With the standardization of space rescue systems resulting from proactive measures that were initiated from international law, the concept of standardization of space systems could be applied to other areas of space equipment and operations.[199] The costs of Inner Solar System infrastructure development are at the phenomenal levels[200] without duplicating and otherwise adding costs by unnecessary international complications due to lack of standards.[201]

---

[1] The Agreement on the Rescue of Astronauts, the Return of Astronauts, and Objects Launhed into Outer Space of 1968. 19 U.S.T. 7570, T.I.A.S. 6569.

[2] The Agreement of the Rescue of Astronauts, the Return of Astronauts, and Objects Launched into Outer Space of 1968. 19 U.S.T. 7570, T.I.A.S. 6569.

[3] Multilateral Treaty on Principles Governing the Activities of States in the Exploration and Use of Outer Space, Including the Moon and other Celestial Bodies, 18 U.S.T. 2410; T.I.A.S. 6347.

[4] The Agreement on the Rescue of Astronauts, the Return of Astronauts, and Objects Launched into Outer Space of 1968. 19 U.S.T. 7570, T.I.A.S. 6569.

[5] The Agreement on the Rescue of Astronauts, the Return of Astronauts, and Objects Launched into Outer Space of 1968. 19 U.S.T. 7570, T.I.A.S. 6569.

[6] Multilateral Treaty on Principles Governing the Activities of States in the Exploration and Use of Outer Space, Including the Moon and other Celestial Bodies, 18 U.S.T. 2410; T.I.A.S. 6347.

[7] Goldman, Nathan. "Transition or Confusion in the Law of Outer Space, in International Space Policy." Legal. Economic and Strategic Options for the Twentieth Century and Beyond. 1987, page 157.

[8] United States of America, Senate: Committee on Foreign Relations. "Hearings on the Treaty on Outer Space. " Congress 1st Session, 1967 at 27 (USA position)

[9] The Agreement on the Rescue of Astronauts, the Return of Astronauts, and Objects Launched into Outer Space of 1968. 19 U.S.T. 7570, T.U.A.S. 6569.

[10] Heath, Gloria W. Space Safety and Rescue 1982-1983. American Astronautical Society. Volume 58, San Diego. Univelt, 1984.

[11] DeSaussure, Hamilton. "Astronaut and Seaman- A Legal comparison." Journal of Space Law. Volume 10, Number 2, 1982, pages 165-179.

[12] The Agreement on the Rescue of Astronauts, the Return of Astronauts, and Objects Launched into Outer Space of 1968. 19 U.S.T 7570, T.I.A.S. 6569.

[13] DeSaussure, Hamilton. "Astronaut and Seaman- A Legal comparison." Journal of Space Law. Volume 10, Number 2, 1982, pages 165-179.

[14] DeSaussure, Hamilton. "Astronaut and Seaman- A Legal comparison." Journal of Space Law. Volume 10, Number 2, 1982, pages 165-179.

[15] The Agreement on the Rescue of Astronauts, the Return of Astronauts, and Objects Launched into Outer Space of 1968. 19 U.S.T 7570, T.I.A.S. 6569.

[16] Stearns, E. V. "Ad Hoc Working Panel on Rescue and Escape." Proceedings of the

Symposium on Space Rendezvous Rescue and Recovery, 1963, pages 220-222.

[17] Mann, Paul. "Reflections on the Cold War: End of the Game, or End of an Inning." Aviation Week and Space Technology, December 18, 1989, pages 18.

[18] Skolnick, Alfred. "The Navy's Final Frontier." Proceedings of the U.S. Naval Institute, January, 1989, pages 29.

[19] Chamberland, Dennis. "Splashdown." Proceedings of the U.S. Naval Institute, January 1989, pages 37-43.

[20] Rao, U.R.; Singh, J.P. Rajan, Y.S. "Earth Safety and Disaster Response Employing Space Borne-Systems – A Review." Acta Astronautica. Volume 18, 1988, pages 347-360.

[21] Multilateral Treaty on Principles Governing the Activities of States in the Exploration and Use of Outer Space, Including the moon and other Celestial Bodies, 18 U.S.T. 2410; T.I.A.S. 6347.

[22] The Agreement on the Rescue of Astronauts, the Return of Astronauts, and Objects Launched into Outer Space of 1968. 19 U.S.T. 7570, T.I.A.S. 6569.

[23] The Agreement on the Rescue of Astronauts, the Return of Astronauts, and Objects Launched into Outer Space of 1968. 19 U.S.T. 7570, T.I.A.S. 6569.

[24] Pepin, E. "Legal Problems Created by the Spautnick." Legal Problems of Space Exploration. A Symposium. Library of Congress, Washington, 1961, 65.

[25] Halsell, James D. Widhalm, Joseph W.; Whitsett, Charles E. "Design of an Interim Space Rescue Ferry Vehicle." Journal of Spacecraft and Rockets. Volume 25, Number 2, March 1988, page 180.

[26] Halsell, James D. Widhalm, Joseph W.; Whitsett, Charles E. "Design of an Interim Space Rescue Ferry Vehicle." Journal of Spacecraft and Rockets. Volume 25, Number 2, March 1988, page 180.

[27] Brown, George. "International Cooperation in Space: Enhancing the World's Common Security." Space Policy. Volume 3, Number 3, August 1987, 1987, pages 166-174.

[28] Mcidull, John C. "Safety Awareness Continuity in Transportation and Space Systems." Acta Astronautica. Volume 17, Number 8, August 1988, pages 931-937.

[29] Garshnek, V. "Crucial Factor: Human: Safety Extending the Human Presnce in Space." Space Policy. August 1989, pages 201-216, especially pages 216,

[30] Hempsell, C. M.; Hannigan, Russell J. "Multi role Capsule System Description." Journal of the British Interplanetary Society. Volume 42, Number 2, February 1989, pages 67-78.

[31] Multilateral Treaty on Principles Governing the Activities of States in the Exploration and Use of Outer Space, Including the moon and other Celestial Bodies, 18 U.S.T. 2410; T.I.A.S. 6347.

[32] Multilateral Treaty on Principles Governing the Activities of States in the Exploration and Use of Outer Space, Including the moon and other Celestial Bodies, 18 U.S.T. 2410; T.I.A.S. 6347.

[33] Christol, C.Q. The International Law of Outer Space. "Chapter 5: The Rescue and Return Agreement." Pages 152-212.

[34] Reynolds, Glen H.; Merges, Robert P. "Rescue and Return of Astronauts." Outer Space: Problems of Law and Policy. Westview Press, Boulder, Colorado, 1989. Pages 194-195.

[35] Goldman, Nathan. "Transition or Confusion in the Law of Outer Space, in International Space Policy." Legal. Economic and Strategic Options for the Twentieth Century and Beyond. 1987, page 157.

[36] Danilenko, Gennady M. "International Law-Making for Outer Space." Space Policy. Volume 5, Number 4, November 1989, pages 321-329.

[37] The Agreement on the Rescue of Astronauts, the Return of Astronauts, and Objects Launched into Outer Space of 1968. 19 U.S.T. 7570, T.I.A.S. 6569.

[38] Multilateral Treaty on Principles Governing the Activities of States in the Exploration and Use of Outer Space, Including the moon and other Celestial Bodies, 18 U.S.T. 2410; T.I.A.S. 6347.

[39] Chamberland, Dennis. "Splashdown." Proceedings of the U.S. Naval Institute, January 1989, pages 37-43.

[40] Article V, Multilateral Treaty on Principles Governing the Activities of States in Exploration and Use of Outer Space, Including the Moon and other Celestial Bodies, 18 U.S.T. 2410; T.I.A.S. 6347.

[41] Article V, Multilateral Treaty on Principles Governing the Activities of States in Exploration and Use of Outer Space, Including the Moon and other Celestial Bodies, 18 U.S.T. 2410; T.I.A.S. 6347.

[42] The Agreement on the Rescue of Astronauts, the Return of Astronauts, and Objects Launched into Outer Space of 1968. 19 U.S.T. 7570, T.I.A.S. 6569.

[43] Multilateral Treaty on Principles Governing the Activities of States in the Exploration and Use of Outer Space, Including the moon and other Celestial Bodies, 18 U.S.T. 2410; T.I.A.S. 6347.

[44] Article V, Multilateral Treaty on Principles Governing the Activities of States in Exploration and Use of Outer Space, Including the Moon and other Celestial Bodies, 18 U.S.T. 2410; T.I.A.S. 6347.

[45] The Agreement on the Rescue of Astronauts, the Return of Astronauts, and Objects Launched into Outer Space of 1968. 19 U.S.T. 7570, T.I.A.S. 6569.

[46] Article V, Multilateral Treaty on Principles Governing the Activities of States in Exploration and Use of Outer Space, Including the Moon and other Celestial Bodies, 18 U.S.T. 2410; T.I.A.S. 6347.

[47] Article V, Multilateral Treaty on Principles Governing the Activities of States in Exploration and Use of Outer Space, Including the Moon and other Celestial Bodies, 18 U.S.T. 2410; T.I.A.S. 6347.

[48] The Agreement on the Rescue of Astronauts, the Return of Astronauts, and Objects Launched into Outer Space of 1968. 19 U.S.T. 7570, T.I.A.S. 6569.

[49] Article V, Multilateral Treaty on Principles Governing the Activities of States in Exploration and Use of Outer Space, Including the Moon and other Celestial Bodies, 18 U.S.T. 2410; T.I.A.S. 6347.

[50] The Agreement on the Rescue of Astronauts, the Return of Astronauts, and Objects Launched into Outer Space of 1968. 19 U.S.T. 7570, T.I.A.S. 6569.

[51] The Agreement on the Rescue of Astronauts, the Return of Astronauts, and Objects Launched into Outer Space of 1968. 19 U.S.T. 7570, T.I.A.S. 6569.

[52] The Agreement on the Rescue of Astronauts, the Return of Astronauts, and Objects Launched into Outer Space of 1968. 19 U.S.T. 7570, T.I.A.S. 6569.

[53] The Agreement on the Rescue of Astronauts, the Return of Astronauts, and

Objects Launched into Outer Space of 1968. 19 U.S.T. 7570, T.I.A.S. 6569.

[54] The Agreement on the Rescue of Astronauts, the Return of Astronauts, and Objects Launched into Outer Space of 1968. 19 U.S.T. 7570, T.I.A.S. 6569.

[55] The Agreement on the Rescue of Astronauts, the Return of Astronauts, and Objects Launched into Outer Space of 1968. 19 U.S.T. 7570, T.I.A.S. 6569.

[56] Article V, Multilateral Treaty on Principles Governing the Activities of States in Exploration and Use of Outer Space, Including the Moon and other Celestial Bodies, 18 U.S.T. 2410; T.I.A.S. 6347.

[57] The Agreement on the Rescue of Astronauts, the Return of Astronauts, and Objects Launched into Outer Space of 1968. 19 U.S.T. 7570, T.I.A.S. 6569.

[58] The Agreement on the Rescue of Astronauts, the Return of Astronauts, and Objects Launched into Outer Space of 1968. 19 U.S.T. 7570, T.I.A.S. 6569.

[59] Article 1 to the The Agreement on the Rescue of Astronauts, the Return of Astronauts, and Objects Launched into Outer Space of 1968. 19 U.S.T. 7570, T.I.A.S. 6569.

[60] The Agreement on the Rescue of Astronauts, the Return of Astronauts, and Objects Launched into Outer Space of 1968. 19 U.S.T. 7570, T.I.A.S. 6569.

[61] The Agreement on the Rescue of Astronauts, the Return of Astronauts, and Objects Launched into Outer Space of 1968. 19 U.S.T. 7570, T.I.A.S. 6569.

[62] Chamberland, Dennis. "Splashdown." Proceedings of the U.S. Naval Institute, January 1989, pages 41.

[63] "Soyuz crew lands safely after aborting reentry twice: Vehicle, crew landed within miles of originally planned site." Aviation Week and Space Technology, Volume 129, Number 11, September 1988, pages 27-29.

[64] Berger, H. "Space Vehicles and Astronaut Assistance and Liability for damage." Proceedings of the 6th Colloquium on the Law of Outer Space. 1966, page 10.

[65] Hosenhall, S. Neil. "Current Issues of Space law before the United Nations." Journal of Space Law, Volume 2, Number 7, 1974, page 7.

[66] The Agreement on the Rescue of Astronauts, the Return of Astronauts, and Objects Launched into Outer Space of 1968. 19 U.S.T. 7570, T.I.A.S. 6569.

---

[67] The Agreement on the Rescue of Astronauts, the Return of Astronauts, and Objects Launched into Outer Space of 1968. 19 U.S.T. 7570, T.I.A.S. 6569.

[68] The Agreement on the Rescue of Astronauts, the Return of Astronauts, and Objects Launched into Outer Space of 1968. 19 U.S.T. 7570, T.I.A.S. 6569.

[69] The Agreement on the Rescue of Astronauts, the Return of Astronauts, and Objects Launched into Outer Space of 1968. 19 U.S.T. 7570, T.I.A.S. 6569.

[70] Article Number 3 to the The Agreement on the Rescue of Astronauts, and Objects Launched into Outer Space of 1968. 19 U.S.T. 7570, T.I.A.S. 6569.

[71] Bogolasky, Jose Claudio. "Agreement between the Government of the Republic of Chile and the Government of the US concerning the use of the Mataveri airport, Easter Island, as a Space shuttle emergency landing and rescue site: A Report." Journal of Space Law. Volume 14, Number 2, 1986, pages 154-159.

[72] The Agreement on the Rescue of Astronauts, the Return of Astronauts, and Objects Launched into Outer Space of 1968. 19 U.S.T. 7570, T.I.A.S. 6569.

[73] The Agreement on the Rescue of Astronauts, the Return of Astronauts, and Objects Launched into Outer Space of 1968. 19 U.S.T. 7570, T.I.A.S. 6569.

[74] The Agreement on the Rescue of Astronauts, the Return of Astronauts, and Objects Launched into Outer Space of 1968. 19 U.S.T. 7570, T.I.A.S. 6569.

[75] Multilateral Treaty on Principles Governing the Activities of States in Exploration and Use of Outer Space, Including the Moon and other Celestial Bodies, 18 U.S.T. 2410; T.I.A.S. 6347.

[76] The Agreement on the Rescue of Astronauts, the Return of Astronauts, and Objects Launched into Outer Space of 1968. 19 U.S.T. 7570, T.I.A.S. 6569.

[77] The Agreement on the Rescue of Astronauts, the Return of Astronauts, and Objects Launched into Outer Space of 1968. 19 U.S.T. 7570, T.I.A.S. 6569.

[78] The Agreement on the Rescue of Astronauts, the Return of Astronauts, and Objects Launched into Outer Space of 1968. 19 U.S.T. 7570, T.I.A.S. 6569.

[79] Agreement Among the Government of the United States of America, Governments of Member States of European Space Agency, the Government of Japan, and the Government of Canada on the Cooperation in the Detailed Design, Development Operation, and Utilization of the Permanently Manned Civil Space

Station.

[80] "Station Partners agree to keep facility design compatible." Space news. April 9-15, 1990, page 12.

[81] "Station Partners agree to keep facility design compatible." Space news. April 9-15, 1990, page 12.

[82] Multilateral Treaty on Principles Governing the Activities of States in Exploration and Use of Outer Space, Including the Moon and other Celestial Bodies, 18 U.S.T. 2410; T.I.A.S. 6347.

[83] The Agreement on the Rescue of Astronauts, the Return of Astronauts, and Objects Launched into Outer Space of 1968. 19 U.S.T. 7570, T.I.A.S. 6569.

[84] The Agreement on the Rescue of Astronauts, the Return of Astronauts, and Objects Launched into Outer Space of 1968. 19 U.S.T. 7570, T.I.A.S. 6569.

[85] The Agreement on the Rescue of Astronauts, the Return of Astronauts, and Objects Launched into Outer Space of 1968. 19 U.S.T. 7570, T.I.A.S. 6569.

[86] Christol, C.Q. The International Law of Outer Space. "Chapter 5: The Rescue and Return Agreement, Conclusion." At Pages 203.

[87] Steven Young, "Society at Astronaut Conference," SpaceFlight Volume 32 (1990) January at page 25.

[88] Steven Young, "Society at Astronaut Conference," SpaceFlight Volume 32 (1990) January at page 25&26.

[89] Chandler, P.P. "US-Soviet intergovernmental agreement on cooperative space activities." Space Policy. Volume 2, Number 1, February 1986, pages 28-36.

[90] Steven Young, "Society at Astronaut Conference," SpaceFlight Volume 32 (1990) January at page 25&26.

[91] Chandler, P.P. "US-Soviet intergovernmental agreement on cooperative space activities." Space Policy. Volume 2, Number 1, February 1986, pages 28-36.

[92] The Agreement on the Rescue of Astronauts, the Return of Astronauts, and Objects Launched into Outer Space of 1968. 19 U.S.T. 7570, T.I.A.S. 6569.

[93] Keesay, Lori. "Robotic System planned to repair orbiting satellites." Space News. December 11, 1989, page 46.

[94] Karen Boehler, "Lifeboat to Safer Shores," Ad Astra (1989), March, at page 9.

[95] Martin, James A. "Integrated Launch and Emergency Entry Vehicle Concept." Journal of Spacecraft and Rockets. Volume 26, Number 5, September-October, 1989, pages 391-392.

[96] Halsell, James D.; Widhalm, Joseph W.; Whitsett, Charles E. "Design of an Interim Space Rescue Ferry Vehicle." Journal of Spacecraft and Rockets. Volume 25, Number 2, March 1988, page 181.

[97] Bogolasky, Jose Claudio. "Agreement between the Government of the Republic of Chile and the Government of the US concerning the use of the Mataveri airport, Easter Island, as a Space shuttle emergency landing and rescue site: A Report." Journal of Space Law. Volume 14, Number 2, 1986, pages 154-159.

[98] "Second Soviet Space Walk expands testing of Spacebikes capability." Space News. February 12-18, 1990, page 10.

[99] The Agreement on the Rescue of Astronauts, the Return of Astronauts, and Objects Launched into Outer Space of 1968. 19 U.S.T. 7570, T.I.A.S. 6569.

[100] The Agreement on the Rescue of Astronauts, the Return of Astronauts, and Objects Launched into Outer Space of 1968. 19 U.S.T. 7570, T.I.A.S. 6569.

[101] Banks, Peter M., Ride, Sally K. "Soviets in Space." Exploring Space: Special Issue. Scientific America. Volume 2, Number 1, 1990, pages 48-55.

[102] "Soviets lead World in Station experience. Space News. January 22-28, 1990, page 14.

[103] Banks, Peter M., Ride, Sally K. "Soviets in Space." Exploring Space: Special Issue. Scientific America. Volume 2, Number 1, 1990, pages 48-55.

[104] "Space Shuttle flights rate and utilization." Space Policy. Volume 3, Number 1, February 1987 pages 5-9.

[105] Brown, N.E. "Space Shuttle Crew Provisions." Space Rescue and Safety Proceedings of the 8th International Space Rescue and Safety Symposium, Volume 4, American Astronautical Society Publications, San Diego, 1976.

[106] "Soviet Test Space Motocycle at MIR." Space News. February 5-11, 1990 page 1&28.

[107] "Second Soviet spacewalk expands testing of spacebike's capability." Space news. February 12-18.1990, page 10.

[108] Bekey, Ivan. "Potential Directions for a Second Generation Space shuttle." Acta Astronautica. Volume 17, Number 9, September 1988, pages 943-952.

[109] Vedda, Jim. Personal Communication. April, 1990.

[110] "Astronaut Candidates Training in China for Future Missions." Aviation Week and Space Technology. February 4, 1980, page 57.

[111] Aviation Week and Space Technology. May 28, 1979, page 26,

[112] Lawlor, Andrew. "Europe, Japan, and Canada upset at possible Space Station changes." Space News. Space News. September 18, 1989, page 42.

[113] The Agreement on the Rescue of Astronauts, the Return of Astronauts, and Objects Launched into Outer Space of 1968. 19 U.S.T. 7570, T.I.A.S. 6569.

[114] The Agreement on the Rescue of Astronauts, the Return of Astronauts, and Objects Launched into Outer Space of 1968. 19 U.S.T. 7570, T.I.A.S. 6569.

[115] National Commission on Space (NCoS). Pioneering the Space Frontier. New York: Bantam, 1986, 211 pages.

[116] National Aeronautics and Space Administration (NASA). Office of Exploration. Beyond Earth's Boundaries: Human Exploration of the Solar System in the 21st Century. 1988 Annual Report to the Administrator. Washington, D.C., U.S. Government Printing Office, 1988, 51pp.

[117] "U.K. plan for space rescue Capsule." Space Flight. Volume 30, January, 1988, page 9.

[118] "U.K. plan for space rescue Capsule." Space Flight. Volume 30, January, 1988, page 9.

[119] The Agreement on the Rescue of Astronauts, the Return of Astronauts, and Objects Launched into Outer Space of 1968. 19 U.S.T. 7570, T.I.A.S. 6569.

[120] The Agreement on the Rescue of Astronauts, the Return of Astronauts, and Objects Launched into Outer Space of 1968. 19 U.S.T. 7570, T.I.A.S. 6569.

[121] Hempsell, C.M. "An extended Space Infrastructure." Journal of the British Interplanetary Society. Volume 42, 1989, Number 1, pages 521-532.

[122] Hempsell, C.M. “An extended Space Infrastructure.” Journal of the British Interplanetary Society. Volume 42, 1989, Number 1, pages 521-532.

[123] Holloway, P.F.; Zerson, W.F.H. “Future Spacecraft Transportation Options-Overview.” American Institute of Aeronautics and Astronautics. paper 86-1210, June 1986.

[124] Hempsell, C.M. “An extended Space Infrastructure.” Journal of the British Interplanetary Society. Volume 42, 1989, Number 1, pages 521&532.

[125] Hempsell, C.M. “An extended Space Infrastructure.” Journal of the British Interplanetary Society. Volume 42, 1989, Number 1, pages 521&532.

[126]Meyer, T.R.; McKay, C.P. “The Resources of Mars Human Settlement.” The Journal of the British Interplanetary Society. Volume 42, Number 4, April 1989, pages 147-160.

[127] The Agreement on the Rescue of Astronauts, the Return of Astronauts, and Objects Launched into Outer Space of 1968. 19 U.S.T. 7570, T.I.A.S. 6569.

[128] Multilateral Treaty on Principles Governing the Activities of States in Exploration and Use of Outer Space, Including the Moon and other Celestial Bodies, 18 U.S.T. 2410; T.I.A.S. 6347.

[129] Pohehleman, Frank. “A concept for a future Autonomous European Space Station.” Acta Astronautica. Volume 18, 1988, pages 385-393.

[130] “Ejectable seats winner for Hermes.” Space News. January 15-21, 1990, page 1.

[131] Pohehleman, Frank. “A concept for a future Autonomous European Space Station.” Acta Astronautica. Volume 18, 1988, pages 391.

[132] Pohehleman, Frank. “A concept for a future Autonomous European Space Station.” Acta Astronautica. Volume 18, 1988, pages 392.

[133] “Station Partners agree to keep facility design compatible.” Space News. April 9-15, 1990., page 12.

[134] Multilateral Treaty on Principles Governing the Activities of States in Exploration and Use of Outer Space, Including the Moon and other Celestial Bodies, 18 U.S.T. 2410; T.I.A.S. 6347.

[135] Gaggero, Eduardo D. “Developing Countries and Space: From awareness to

participation." Space Policy. Volume 5, Number 2, 1989, pages 107-110.

136 Sorenson, Pal. "Development of Norwegian Space Activities." Journal of the British Interplanetary Society. Volume 43, Number 3, March 1990, pages 91-92.

137 Sorenson, Pal. "Development of Norwegian Space Activities." Journal of the British Interplanetary Society. Volume 43, Number 3, March 1990, pages 91-92.

138 Anderson, B.N. "Space Research in Norway." Journal of the British Interplanetary Society. Volume 43, Number 3, March 1990, page 127.

139 Sorenson, Pal. "Development of Norwegian Space Activities." Journal of the British Interplanetary Society. Volume 43, Number 3, March 1990, pages 91-92.

140 Marcus, Daniel. "Iraqis claim Launch Success." Space News. December 11, 1989, page 16.

141 Simpson, John; Acton, Philip; Crowe, Simon. "The Israeli satellite Launch: Capabilities, intentions and implications." Space Policy. Volume 5, Number 2, May 1989, pages 117-128.

142 "Israeli Satellite Launch." Janes Defense Weekly. September 24, 1988, pages 703.

143 The Agreement on the Rescue of Astronauts, the Return of Astronauts, and Objects Launched into Outer Space of 1968. 19 U.S.T. 7570, T.I.A.S. 6569.

144 The Agreement on the Rescue of Astronauts, the Return of Astronauts, and Objects Launched into Outer Space of 1968. 19 U.S.T. 7570, T.I.A.S. 6569.

145 The Agreement on the Rescue of Astronauts, the Return of Astronauts, and Objects Launched into Outer Space of 1968. 19 U.S.T. 7570, T.I.A.S. 6569.

146 Clarke, Arthur C. Space Odyssey: 2001.

147 "Soyuz crew lands safely after aborting re-entry twice: Vehicle crew landed within miles of originally planned site." Aviation Week and Space Technology. Volume 129, Number 11, September 12, 1988, page 27-28.

148 Lawlor, Andrew. "Cosmonauts dock with MIR after automatic systems fails." Space News. Space News. September 18, 1989, page 41.

149 Halsell, James D. Widhalm, Joseph W.; Whitsett, Charles E. "Design of an Interim Space Rescue Ferry Vehicle." Journal of Spacecraft and Rockets. Volume 25, Number 2, March 1988, pages 180-186.

[150] The Agreement on the Rescue of Astronauts, the Return of Astronauts, and Objects Launched into Outer Space of 1968. 19 U.S.T. 7570, T.I.A.S. 6569.

[151] Gabrynowicz, Joanne I. "Space Law: A Case for the Cosmos." Space World. December, 1988, pages 7-9.

[152] Multilateral Treaty on Principles Governing the Activities of States in Exploration and Use of Outer Space, Including the Moon and other Celestial Bodies, 18 U.S.T. 2410; T.I.A.S. 6347.

[153] Gabrynowicz, Joanne I. "Space Law: A Case for the Cosmos." Space World. December, 1988, pages 7-9.

[154] Gabrynowicz, Joanne I. "Space Law: A Case for the Cosmos." Space World. December, 1988, pages 7-9.

[155] Gabrynowicz, Joanne I. "Space Law: A Case for the Cosmos." Space World. December, 1988, pages 7-9.

[156] The Agreement on the Rescue of Astronauts, the Return of Astronauts, and Objects Launched into Outer Space of 1968. 19 U.S.T. 7570, T.I.A.S. 6569.

[157] The Agreement on the Rescue of Astronauts, the Return of Astronauts, and Objects Launched into Outer Space of 1968. 19 U.S.T. 7570, T.I.A.S. 6569.

[158] Cousins, D.; Akin, D.L. "Moments Applied in the Rotation of the Massive Objects in Shuttle Extravehicular Activity." Journal of Spacecraft and Rockets. Volume 26, Number 4, July-August, 1989, pages 293-294.

[159] Cook, Karla. "NASA considers design, cost issues for Freedom era spacesuits." Space News. September 18, 1989, page 45.

[160] Kennedy, Gregory P. "Jet Shoes and Rocket Packs: The Development of Astronaut Manoeuvring Units." Space World. October 1984, 4-9.

[161] Von Kries, Wulf. "Flunking on Space Station Cooperation." Space Policy. Febuary 1987, pages 10-12.

[162] "Station partners agree to keep facility design." Space News. April 9-15, page 12.

[163] Schnell, M. Ph.D Personal Oral Communication, 1989.

[164] The Agreement on the Rescue of Astronauts, the Return of Astronauts, and Objects Launched into Outer Space of 1968. 19 U.S.T. 7570, T.I.A.S. 6569.

[165] "International docking standards, Hermes Station Rescue studies." Aviation Week and Space Technology. September 19, 1988, page 9.

[166] The Agreement on the Rescue of Astronauts, the Return of Astronauts, and Objects Launched into Outer Space of 1968. 19 U.S.T. 7570, T.I.A.S. 6569.

[167] Ambrus, Judith H.; Herman, Daniel H. "The Impact of Launch Vehicle Constraints on U.S. Space Station Design and Operations." Acta Astronautica. Volume 18, 1988, page 45.

[168] Couvalt, Craig. "Ice Clouds could threaten Space Shuttle Reentry safety." Aviation Week and Space Technology. August 22, 1988, pages 83.

[169] The Agreement on the Rescue of Astronauts, the Return of Astronauts, and Objects Launched into Outer Space of 1968. 19 U.S.T. 7570, T.I.A.S. 6569.

[170] The Agreement on the Rescue of Astronauts, the Return of Astronauts, and Objects Launched into Outer Space of 1968. 19 U.S.T. 7570, T.I.A.S. 6569.

[171] Gabrynowicz, Joanne I. Personal Communication, 1990.

[172] Reinalt, Michael. (ret) German Airforce. Personal Oral Communication, April 1990.

[173] Reinalt, Michael. (ret) German Airforce. Personal Oral Communication, April 1990.

[174] Qizhi, He. "On Strengthening the Role of COPUOS: Maintaining Outer Space for Peaceful Uses." Space Policy. Volume 2, Number 1 , February 1986, pages 3-6.

[175] Von Kries, Wulf. "Flunking on Space Station Cooperation." Space Policy. February 1987, pages 10-12.

[176] The Agreement on the Rescue of Astronauts, the Return of Astronauts, and Objects Launched into Outer Space of 1968. 19 U.S.T. 7570, T.I.A.S. 6569.

[177] The Agreement on the Rescue of Astronauts, the Return of Astronauts, and Objects Launched into Outer Space of 1968. 19 U.S.T. 7570, T.I.A.S. 6569.

[178] Dembling, Paul G.; and Arons, Daniel M. "The Evolution of the Outer Space Treaty." Journal of Air Law and Commerce. Volume 33, 1967, page 436.

[179] The Agreement on the Rescue of Astronauts, the Return of Astronauts, and Objects Launched into Outer Space of 1968. 19 U.S.T. 7570, T.I.A.S. 6569.

[180] Lum, H.; Heer, E. "Intelligent, Autonomous Systems in Space." Acta Astronatica. Volume 17, Number 10, October 1988, pages 1081-1092.

[181] Balenbov, V.M.; Zubenko, G.I.; Voronov, D.A.; Valnicek, B.; Rechek, I.; Rujichka, I. "The Flight Telerobotic Servicer (FTS): A Focus for Automation and Robotics on the Space Station." Acta Astronautica. Volume 17, August 1988, pages 759-768.

[182] The Agreement on the Rescue of Astronauts, the Return of Astronauts, and Objects Launched into Outer Space of 1968. 19 U.S.T. 7570, T.I.A.S. 6569.

[183] The Agreement on the Rescue of Astronauts, the Return of Astronauts, and Objects Launched into Outer Space of 1968. 19 U.S.T. 7570, T.I.A.S. 6569.

[184] "NASA advisory panel recommands expensive vehicle plan." Space news. Febuary 12-18, 1990, page 22.

[185] "Crew rescue studies go forward amid scepticism." Space News. January 22-28, 1990, page 16.

[186] "NASA: No permanent station crew without escape vehicle." Space News. February 12-18, 1990, page 6.

[187] The Agreement on the Rescue of Astronauts, the Return of Astronauts, and Objects Launched into Outer Space of 1968. 19 U.S.T. 7570, T.I.A.S. 6569.

[188] The Agreement on the Rescue of Astronauts, the Return of Astronauts, and Objects Launched into Outer Space of 1968. 19 U.S.T. 7570, T.I.A.S. 6569.

[189] Multilateral Treaty on Principles Governing the Activities of States in Exploration and Use of Outer Space, Including the Moon and other Celestial Bodies, 18 U.S.T. 2410; T.I.A.S. 6347.

[190] Christol, C.Q. The International Law of Outer Space. "Chapter 56: The Rescue and Return Agreement." Pages 209.

[191] "International Docking Standards, Hermes Station Rescue Studies." Aviation Week and Space Technology. September 19, 1988, page 9.

[192] Hannigan, Russell J. "Multi-Role Capsule Operations." Journal of the British Interplanetary Society. Volume 42, Number 2, February 1989, page 84-85.

[193] "Design Cost Issues for Freedom-Era Spacesuits." Space News. September 18. 1989, page 45.

---

194 Uy, O.M.; Maurer, R.H. "Quality Assurance Requirements for a Large Li/$SoCl_2$ Battery for Spacecraft Applications." Journal of Spacecraft and Rockets. Volume 25, Number 4, July-August, 1988, pages 304-308.

195 National Aeronautic and Space Administration. Manned Manoeuvring Unit (MMU) Operational Data Book, Volume 1, NASA contract NA S9-17018, Denver Aerospace Division, Martin Marietta Aerospace Corporation, July 1985.

196 National Aeronautic and Space Administration. Manned Manoeuvring Unit (MMU) Operational User's Guide, Volume 1, NASA contract NA S9-17018, Denver Aerospace Division, Martin Marietta Aerospace Corporation, January 1984.

197 Halsell, James D.; Widhalm, Joseph W.;& Whitsett, Charles E. "Design of an Interim Space Rescue Ferry Vehicle." Journal of Spacecraft and Rockets. Volume 25, Number 2, March-April, 1988, pages 180, 186.

198 Chandler, P.P. "US-Soviet Intergovernmental Agreement on Cooperative Space Activities: Should it be Reestablished." Space Policy. Volume 2, Number 1, February 1986, pages 28-36.

199 Serafimov, K.B. "Achieving Worldwide Cooperation in Space." Space Policy. Volume 5, Number 2, May 1989, pages 111-116.

200 Hempsell, C.M. "An Extended Space Infrastructure." Journal of the British Interplanetary Society. Volume 42, 1989, Number 1, page 521.

201 "International Docking Standards, Hermes Station Rescue Studied." Aviation Week and Space Technology. September 19, 1988, page 9.

www.ingramcontent.com/pod-product-compliance
Lightning Source LLC
Chambersburg PA
CBHW021839020426
42334CB00014B/700

* 9 7 8 1 8 9 7 4 7 2 2 1 7 *